FOOD & FARMING

IN HOT AND COLD PLACES

MALCOLM PENNY

Wayland

TITLES IN THIS SERIES

Animals in Hot and Cold Places
Clothes in Hot and Cold Places
Food and Farming in Hot and Cold Places
Homes in Hot and Cold Places

Series editor: Geraldine Purcell
Series designer: Helen White

Cover: (top) This Saami woman is helping to herd reindeer in northern Norway. (bottom) Rice grows well in the tropical climate of Sri Lanka.

Title page: Crops do not grow in the Arctic region so people rely on hunting and fishing for their food supply. Here, cod are being dried on Lofoten Island, in Norway.

First published in 1994 by Wayland (Publishers) Limited
61 Western Road, Hove, East Sussex BN3 1JD, England

British Library Cataloguing in Publication Data
Penny, Malcolm
Food and Farming in Hot and Cold Places. – (Hot & Cold Places)
I. Title II. Price, David III. Series 338.1

ISBN 0 7502 0808 2

Typeset by White Design
Printed and bound in Great Britain by BPC Paulton Books Ltd., Paulton

CONTENTS

THE FIRST FARMERS

▼ **This picture shows an Egyptian farmer harvesting wheat in 1250 BC. The growing of crops, such as wheat and barley, had spread from the part of Asia known as the Middle East.**

Once, there were no farmers. Instead, people hunted wild animals and gathered whatever food they could find as they wandered about. We call these people 'hunter-gatherers'.

In the Kalahari Desert, in southern Africa, the Bushmen still hunt animals with spears, or bows and arrows. The women do not go hunting – they are responsible for finding fruits and roots to cook with the meat. Once, all humans were nomadic hunter-gatherers. Later, they began to change their ways and settled down to become farmers.

The change took place about 10,000 years ago, in the Middle East. Some of the first farmers lived in hilly areas in the north of what is now Iran and Iraq.

◀ **These Kalahari Bushmen are taking one or two ostrich eggs from a nest, for food. The Bushmen live in the Kalahari Desert, in Africa. They are one of the few groups of people who are still nomadic hunter-gatherers.**

This area had some rainfall, but not enough to support forests. The hills were covered in grasses – among them wild wheat and barley, which are grasses with big, edible seeds. Grazing wild on the hills were sheep, goats, pigs, cattle and horses. The rain cut rivers in the valleys, so water could be found in holes in the riverbeds even in dry weather.

The people learned to plant grain as well as gathering it. They also domesticated animals so that they did not have to hunt them. Later, people began to build houses out of mud and sticks, instead of temporary shelters made of grass and animal skins. Slowly, over many hundreds of years, the first farming villages grew up.

▼ These Arab children in Syria are looking after a flock of sheep. Thousands of years ago, sheep and goats were some of the first animals to be domesticated by humans in the Near and Middle East.

The first farmers settled and began growing crops because the soil and the growing conditions were very good. In the thousands of years that have passed since these first farming settlements, farmers have learned to grow food and raise animals in many different places with very varying climates.

▲ This farmer and his son, in Iran, are using a wooden sledge to separate the outer shells of the grain from the seeds. This is called threshing.

FARMING IN AFRICA

All crops need the right amount of sunshine and enough water to grow well. The seasons produce problems for farmers in tropical Africa. The climate is divided between a dry season – when nothing will grow – and a wet season, called 'the rains'. If the rains do not come there is a drought and the crops die in the field. A drought can lead to famine – when no food is available and people die from hunger.

▲ These farmers in Mali are breaking up the hard, dry soil with hand-hoes before planting a crop of millet.

However, the farmers have ways of overcoming all but the worst weather. In Kenya where the main crop is maize, farmers often plant another crop as well, in case there is not much rain. The second crop might be a grass with edible seeds, such as cassava, sorghum, or millet. These do not need as much water as maize to grow well.

Some Kenyan farmers plough their land with oxen, but most till (turn over) the soil with a hand-hoe, called a *jembe*. They scatter the seed, instead of planting it in rows. For fertilizer, they use cattle droppings.

▼ This photograph shows a typical small farm in Kenya. There are fields on the hillsides and on the low ground so that the farmer can decide where the best areas to plant will be, depending on the amount of rain expected that year.

WILL IT RAIN?

In Kenya the best small farms, or *shambas*, are partly in a valley and partly on the side of a hill, so that the farmer can choose where to plant the crops. If the farmer thinks there will not be very much rain one year, then it is better to plant on the low ground. This is so that the water runs off the hills and stays on the flatter fields. Too much water may rot the crop. If the farmer expects plenty of rain, then the crop should be planted on the hillside to let the water run off.

There is a danger in planting on a slope. If the rain is too heavy, it might wash the soil away down the hill. To stop this, many farmers make terraces, like wide steps, so that they can plant their crops in horizontal, level strips along the hillside.

'SLASH AND BURN'

Many tropical rainforests are in danger because of the farming activity called 'slash and burn'. Where the ground is covered with trees and bushes, farmers have to clear places to plant their crops. One way of doing this is to chop the bushes down and ring-bark the trees. When everything is dry, the farmers set fire to it. Afterwards, the ground is not only clear but also fertilized with ash from the fire.

▲ These people living in the rainforest of Guatemala use the slash and burn method to clear land to plant their crops.

The trouble with slash and burn farming is that, without fallen leaves to trap water and rot into the ground, the soil soon loses its goodness. After a couple of years, it grows very poor crops.

The farmers then leave the first place and clear another for their crops. When that is used up, they clear another, and so on. Eventually, they return to the first place – now covered with scrub (small trees and bushes) once more. When they clear this area it grows good crops again.

Slash and burn farming is used in places as far apart as Central America and Indonesia. In southern Mexico, people have been farming like this for thousands of

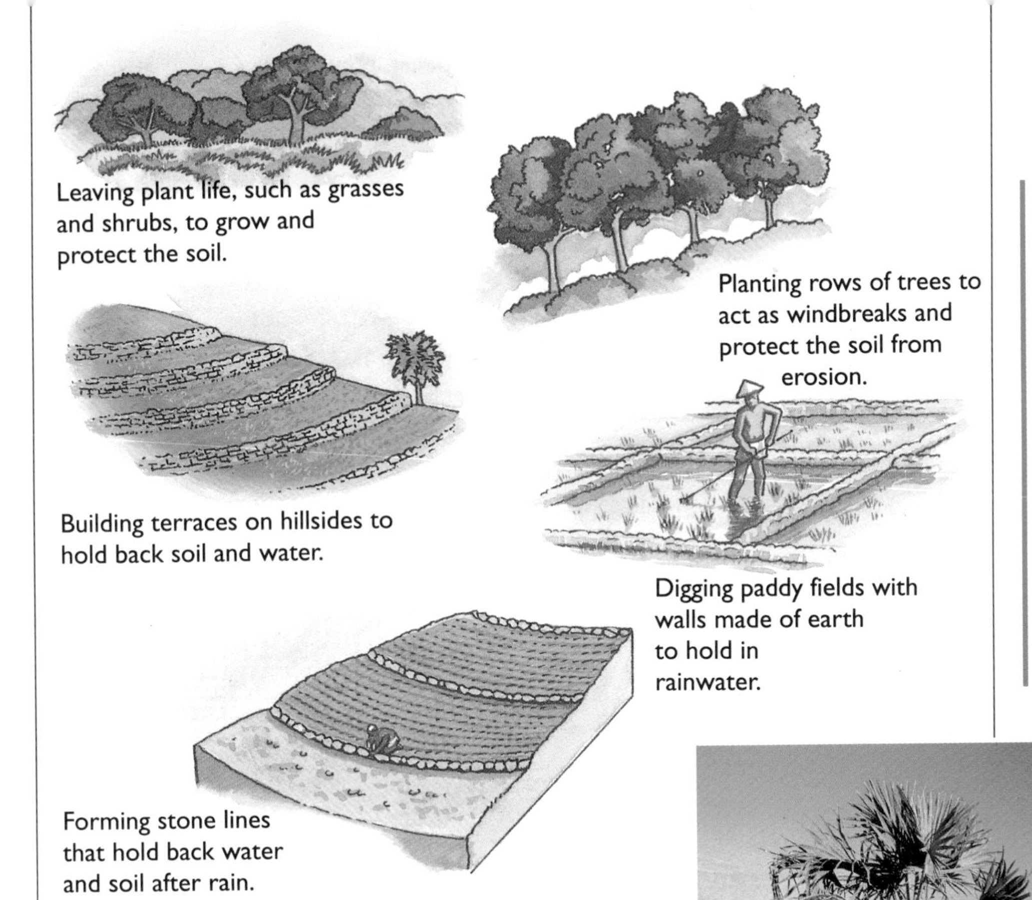

◀ The roots of trees and bushes help to keep soil in place. When trees are cut down the soil is left without protection from wind and rain – this leads to soil erosion. These methods can help to save soil from being blown or washed away.

years and the land still grows good crops. But it works only if there is enough land for each village to have at least five lots of fields, so that every lot can be left for four years to lie fallow before it is cleared again. If there is not enough land – or too many people – slash and burn leads to soil erosion and famine. One of the worst examples of soil erosion caused by slash and burn is on the island of Madagascar, in the Indian Ocean. Farmers on the island, who originally came from Indonesia, found that the forests did not grow back after they had been burned. Rain washed away the soil and so crops could not grow. Now, where forests once stood, there are only bare hills.

▲ This hillside in Madagascar was once covered in thick rainforest. Now, because earlier farmers cut down the forest, only grass and shrubs grow there.

RICE: FLOODED FIELDS

Rice is another type of grass with edible seeds. It is the staple food of more than half the people in the world.

▲ **Planting out rice seedlings (young plants) is very hard work. The seedlings are grown in a dry field. When the paddies have been prepared and are full of water, the seedlings are planted out one at a time.**

Rice gives a good crop only when its roots are underwater, so it needs a warm climate and plenty of rain.

To make sure that the rice is always underwater, rice farmers have to make their own floods. They make paddies, which are pools dug out by hand. They build a wall of earth around each paddy and then dig channels for water from a nearby stream to flow into the paddies.

To prepare the bottom of the paddy for planting, some farmers bring their cattle into it and drive them round and round. This makes sure that the soil is soft enough for the roots of the rice to grow well.

Planting rice is very hard work and it needs a lot of people to help. Each planter carries a bunch of rice seedlings and places each plant into the mud, forming straight rows.

When all the low-lying land is used up, the farmers make paddies on hillsides by building terraces. Each level terrace has an earth wall on the downhill side to keep the water in – except in one place, where the water can flow down to the next terrace.

When the rice is ripe, the farmers drain the water from the paddies to make harvesting easier.

◀ **Rice growing on terraced hillsides surrounding this village in the Philippines.**

▶ A Cree fisherman in northern Canada using a net to catch fish through a hole in the ice.

HUNTERS IN THE SNOW

In very cold places, such as the Arctic and Antarctic regions, it is impossible to grow crops. It is too cold and dry – the land is also covered by ice for most of the year. In the Arctic there are a few wild berries and other plant foods, but only during the short summer. People in the Arctic areas must live as hunter-gatherers. They are very skilled at trapping and hunting animals for food.

During the summer, in northern Canada and Alaska, USA, groups of native Americans make temporary camps near lakes or rivers to catch fish. They use traps or nets to catch the salmon that migrate up the rivers. They used to preserve their fish by smoking or drying them. Today many villages have an electric freezer which keeps the catch fresh.

In late summer these native Americans gather berries. They eat most of them straight away, but they store some to eat in the winter. The berries are important, because they contain vitamin C (which helps to keep them healthy). Some plants can be cooked and eaten in summer, but mostly the people live on meat and fish, either fresh or preserved.

In winter, the native Americans go hunting. Hares and porcupines are good to eat, or sometimes they can catch a fat beaver. Two important food animals are moose and caribou (reindeer). In the past the native Americans would hunt them with spears and axes, but today they use shotguns.

The other way of getting meat is by trapping animals. The native Americans eat or use every bit of the animals they catch. They trap wolves, mink, foxes and otters, using the skin to make warm trousers, and knee-length coats called parkas.

▲ This Cree woman is gathering cranberries in Canada.

THE FROZEN SEA

▶ These Inuit hunters in Greenland have shot a seal at its breathing-hole.

▼ An Inuit fisherman makes a catch through a hole he has cut in the ice.

The Inuit who live in the Arctic on the northern coasts of North America and the shores of Greenland live by hunting and fishing. The summer lasts for only three months, so most of their hunting is done on the frozen sea, in the long, dark months of winter.

Seal hunting in winter needs patience, strength and great skill. The hunter must wait in the dark beside a seal's breathing-hole until the animal comes up for air. A seal can stay underwater for only about ten minutes, but because each one has as many as twenty breathing-holes in its territory, the hunter might have to wait more than three hours for the chance to shoot or harpoon it. The hunter can tell when the seal is coming by watching for the movements of a small rod placed in the top of the hole.

The seal provides the hunter and his family with meat, clothes from its skin, needles and other tools made from its bones, and fuel. Seal fat is good to cook with and it also gives light and heat when it is burned in lamps and stoves.

While the hunter is away, the women in his family set traps for foxes or hares. When the family has eaten the meat, the skins are made into clothes.

HUNTING POLAR BEARS

Hunting polar bears was the most dangerous part of Inuit life in the past. Hunters would track them with dogs and then kill them with spears. During a hunt, dogs could be killed or the hunters wounded by polar bears. Today, bear hunting and seal hunting are easier and safer because the Inuit use shotguns instead of spears and harpoons. Although polar bears are protected from all other hunters, the Inuit are allowed to kill as many as they need for food and skins.

◀ Hunting polar bears is not so dangerous now that the Inuit use shotguns instead of spears.

SAAMI REINDEER

Not all Arctic people are hunter-gatherers like the native Americans and the Inuit. The Saami, or Lapps, who live along the northern coasts of Norway, Sweden and Finland – inside the Arctic Circle – lead a nomadic way of life, following their herds of reindeer through the changing seasons.

In spring they travel north and west, following the herds as they feed on the grass on the hills, which are now free

▼ **The Saami people are sometimes called Lapps. They wear colourful traditional clothing as they follow their migrating reindeer herds.**

HERDERS

▲ **This herder is using a length of cloth to help get the herd into a corral.**

▲ **(ABOVE RIGHT) A Saami summer camp in northern Norway. When winter comes, they will follow their herds south on sledges pulled by reindeer.**

of the snow that covered them during winter. In winter they move south and east, into the forests where the snow is not so deep and the reindeer can feed.

The Saami live in movable tents made of poles and reindeer hides (skins). Each family owns some reindeer, from which they get milk, meat and leather. They use the reindeer antlers to make buttons and carved ornaments.

Some of the Saami make a lot of money from selling reindeer meat for export. A few can afford the luxury of a helicopter or aeroplane to take most of the group straight to the summer feeding grounds, while a few men stay behind to follow the reindeer herd.

FISHING IN TROPICAL WATERS

The shallow water near tropical shores has a good supply of fish. Between coral reefs and the beach, there is often a sandy area which is flooded at high tide. This is where the people usually put their fish traps. Sometimes they put them on top of the reef itself.

The traps vary from place to place. A common kind in the islands of the Seychelles is a basket trap which is

▲ These fishermen are using spears to catch fish as they swim among coral reefs.

◀ Chinese fishing nets are often used in southern India. The nets, which are attached to poles, are lowered into the water. When a shoal of fish swim over the nets, the fishermen raise the nets out of the water with the poles.

▲ This man is fishing from his canoe on Lake Victoria, in Kenya.

made by crisscrossing bamboo. The basket is shaped so that it is easy for fish to swim into, but hard for them to find the way out. The gaps in the crisscross pattern are big enough to let the smallest fish go, because they are not worth eating. The trap is weighted down with heavy stones and it has bait (food) inside to attract the fish.

Just outside the reef, people fish from boats with hand-lines and hooks. They have to be careful what they keep to eat, because some tropical fish are poisonous. Small sharks are a very welcome catch, because they taste delicious. Bigger fish such as groupers and parrot fish make a good meal, too.

To catch large, fast-swimming fish such as caranx, tuna and sailfish, people need to use fast motor boats and strong lines.

CATTLE OR CAMELS ?

▲ Cattle do not do well in very hot, dry climates. This herd of cattle in Mali is trying to find some shade from the hot sun.

▼ The nomadic Masai herders in Kenya are famous for their large herds of cattle.

In Africa, plants that grow in arid (dry) places are often salty, and cattle or goats cannot eat them. The animals become thin and many of them die. If the people milk their hungry cows and goats, the young calves and kids will not have enough to eat. If the people eat the animals, they will have no livestock to breed from when the rains return and the grass grows.

In a drought, this problem causes many people to starve, along with their animals. One answer is to find an animal that will produce food even in dry weather. There is such an animal: the camel.

The camel's favourite food plants contain plenty of water, but are too thorny and salty for cattle and goats to eat. Camels survive better on these desert plants than by eating grass.

If cattle, sheep or goats cannot get enough fresh water, they stop eating after two or three days. Camels can drink salty-tasting water which other animals leave alone. They go on eating, growing and producing milk. The camel's milk is good for people to drink and it keeps better than cows' milk even in warm weather.

It is difficult to keep camels because they do not breed until they are six years old and they can only have a calf every other year. Also, they are hard to look after when they wander in the desert to feed.

Now, scientists have discovered that if the camels are kept in pens instead of being allowed to wander, they breed every year after they are two years old. The scientists are teaching people in the dry areas of Africa how to keep camels instead of cattle, so that they can have milk and meat even in the dry season.

▼ Camels can drink salty water which cattle will not touch.

▶ **This boy in Mozambique, Africa, has climbed to the top of a coconut tree to collect the coconut pods.**

▲ **Coir is collected from the outer husks (shells) of coconuts. It can be twisted into very coarse thread, for making mats.**

TROPICAL PLANTATIONS

COCONUTS

The coconut palms that bend over tropical beaches in travel advertisements were once much more important than they are today. The coconuts that they produce were the source not only of food for local people, but also of valuable products which could be sold all over the world. One of the products was copra, which is the dried flesh of the coconut, and a source of edible oil. The other was coir, the rough fibres from the outside of the nut.

The oil from copra was used not only for food, but also to make soap and candles. Coir can be twisted into a coarse thread and made into mats, called 'coconut matting'.

Coconuts are still grown in plantations, where the palms stand in straight rows on neatly trimmed grass: but the trade is much less than it was. Modern synthetic products are cheaper to make and last longer. There is still a small local trade in coconut products but many of the people who used to work on the plantations no longer have a job.

There are some modern uses for the coconut: coir is used by some gardeners to stop weeds growing on flower beds, and the oil from copra is still used in some soaps.

▼ **Breadfruit is a good source of food.**

BREADFRUIT

There are other tropical trees whose fruit is still an important food. Among them is the breadfruit – a large oval fruit, like a giant potato, which can be cooked by boiling, baking or frying. It grows on large trees which do not need looking after. One tree can provide enough fruit for a whole village.

FOOD FROM AROUND THE WORLD

▶ **This woman is selling fruit and vegetables from a roadside stall in Kenya. Nowadays, produce from tropical countries is often exported to colder countries.**

A tour of a supermarket is a good way to find out what foods are produced and exported by countries with hot or cold climates. Now that so much food is carried by air, even perishable foods from abroad, such as fruit, can be found on the shelves.

Fresh fish from reefs in the Indian Ocean and the Caribbean, and soft fruits from the tropics, such as yams and papaw, are now quite common in supermarkets in colder countries. These foods can now be flown in, to arrive only a day or two after they were picked. Other fruits are much more familiar: bananas, oranges, avocado pears and coconuts have been imported by sea for years.

Foods from cold countries are usually on the fish counter. Cod, salmon and halibut all come from Arctic seas – though they are not usually caught by the people who live there. The Inuit and the native Americans do not have big enough boats or the equipment to catch fish in great numbers.

Some of the fish comes from the Antarctic Ocean: but most of what is caught there goes to Japan and Russia, not to Europe or North America. The main catch in the Antarctic is cod and squid – though krill is also gathered, mostly for animal feed.

◀ Cod drying on racks near a fishing port in Norway.

CASH CROPS

▲ **These women in a market in Mali are selling spare food to earn money to buy other goods.**

The crops that many farmers grow in developing countries are not for their families to eat. They are grown to sell to other countries. These 'cash crops' include, cotton, tea, coffee and cocoa. They are grown on plantations. Local people who work on the plantations use their wages to buy food in the market.

Another way for people in developing countries to earn extra money is to use their own land to grow crops such as bananas, oranges or pineapples. They then sell these – either in the local market, or to a company which exports the fruit. In both cases, the growers then need to buy food for themselves.

For people who live in small villages, the market may be too far away. In this case they grow enough food to eat and exchange any spare crops with their neighbours. The exchange can be for different food, or perhaps for clothes or fuel. This way of trading is called 'barter'.

Sometimes the entire income of a country depends on the sale of cash crops overseas. If a lot of land is used to grow cash crops, there may not be enough left to grow food to eat. In Zambia, maize is the main food crop, but tobacco and tea are more valuable. Land which could be growing maize is used to grow cash crops to sell abroad.

It is a serious problem. If the land is used to grow food for local people, the country will not have enough money for schools and hospitals. If enough cash crops are grown to pay for these essentials, the people may not have enough to eat. The problem is made harder as the population of the country increases, so that more food and more money is needed to look after the increased number of people.

▼ A tea picker in Zambia, in Africa.

GLOSSARY

Antarctic The cold, ice-covered land around the South Pole.
Arctic The area of frozen sea and ice-covered lands around the North Pole.
breed To produce young.
coral reefs Rock-like structures in shallow, tropical water which are made up of the bodies of thousands of tiny sea creatures. Many small sea creatures and fish live on and around the reefs.
domesticated Animals which have been tamed so that humans can use their products, such as meat, milk or wool.
edible Food that can be eaten.
erosion When soil has been washed away by water or blown away by wind.
export A product, such as food, sold by one country to another.
fallow Ploughed land which is left unplanted for a period of time.
fertilizer A substance, such as animal droppings or chemicals, put into the soil to make it better for growing crops.
harpoon To use a harpoon, a long dart, usually with a line attached to it, to catch sea creatures such as seals and whales.
harvesting To pick or collect a crop.
income Money that has been earned by a person or a country.
krill Shrimp-like creatures which live in the ocean.
migrate To travel great distances to find a better food supply or to reach breeding grounds.

nomadic People who do not live in just one place. They travel around to find food and water.
perishable Fresh food which will not keep for long.
preserve To keep food from going bad.
ring-bark To cut through the bark of a tree all the way round, so that the tree dies.
staple food The main type of food that people eat.
synthetic Something which does not come from an animal or plant but is made by humans.
tropical The hot areas of the world that lie between two imaginary lines around the Earth, which we call the Tropic of Cancer and the Tropic of Capricorn.

FURTHER READING

Focus on Kenya by Fleur Ng'weno (Hamish Hamilton, 1990)
Food and Farming (Young Geographer series) by Susan Reed-King (Wayland, 1992)
Food Resources (The World's Resources series) by Robin Kerwood (Wayland, 1993)
Living Arctic by Hugh Brody (Faber & Faber, 1987)

PICTURE ACKNOWLEDGEMENTS

B. & C. Alexander *cover* (top), 8 , 14, 15, 17, 18, 18-19, 19, 22 (top), 28; Eye Ubiquitous 13 (B. Dean), 20 (bottom) (D. Cumming), 24 (inset); Frank Lane Picture Agency 22 (bottom) (P. Davey); Geoscience Features 25; Michael Holford 4; Hutchison Picture Library 26 (inset) (P. Wolmuth); Images Colour Library *cover* (bottom); Images of Africa/(D.K. Jones) 9, 21, 23, 26; Life File 6 (D. Heath), 7 (S. Kay); N.H.P.A. 16 (both) (B. & C. Alexander); Oxford Scientific Films Ltd 5 (A. Bannister), 11 (D. Allan), *title page* and 27 (D. Allan); Tony Stone Worldwide 10 (D. Hiser), 20-21 (D.Hiser), 24 (P. Lamberti), 29 (I. Murphy); Wayland Picture Library 12 (J. Waterlow). Inside artwork by David Price. Cover artwork by William Donohoe.

INDEX

Numbers in **bold** indicate entries which are illustrated.